ODD COUPLE

ODD COUPLE

Susannah Marriott

MQP

Published by MQ Publications Limited
12 The Ivories
6–8 Northampton Street
London, N1 2HY
Tel: +44 (0)20 7359 2244
Fax: +44 (0)20 7359 1616
E-mail: mail@mqpublications.com

North American Office
49 West 24th Street
8th Floor
New York, NY 10010
E-mail: information@mqpublicationsus.com

Web site: www.mqpublications.com

Copyright © 2006 MQ Publications Limited
Text copyright © 2006 Susannah Marriott

ISBN: 1-84601-061-6

123456789

All rights reserved. No part of this publication may be reproduced or transmitted in any form or by any means, electronic or mechanical, including photocopy, recording, or any information storage and retrieval system now known or to be invented without permission in writing from the publishers.

Printed in China

Do one thing every day that scares you.

ELEANOR ROOSEVELT

Whatever our souls are made of, his and mine are the same.

EMILY BRONTË

The critical period in matrimony is breakfast-time.

A.P. HERBERT

> Inside of me there's a thin person screaming to get out.
> —Just the one, dear?
>
> **EDINA AND HER MOTHER**
> *ABSOLUTELY FABULOUS*

Quarrel? Nonsense; we have not quarreled. If one is not to get into a rage sometimes, what is the good of being friends?

GEORGE ELIOT

If God had wanted me otherwise, He would have created me otherwise.

JOHANN VON GOETHE

> Excuse me, but weren't we blissfully married in a past life?
>
> ANONYMOUS

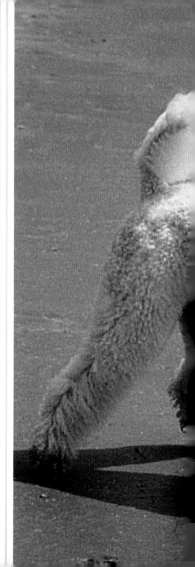

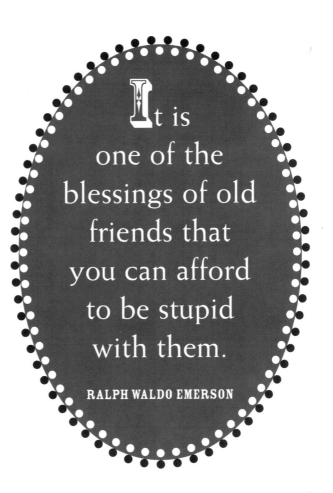

It is one of the blessings of old friends that you can afford to be stupid with them.

RALPH WALDO EMERSON

I exercise strong self control. I never drink anything stronger than gin before breakfast.

W. C. FIELDS

You make me melt like hot fudge on a sundae.

ANONYMOUS

O the
pleasures of
neighborly chat,
If you can but keep
scandal away,
To learn what the
world has been at,
And what the great
Orators say.

ROBERT BLOOMFIELD

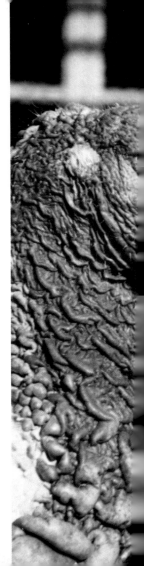

You know that look women get when they want sex? Me neither.

STEVE MARTIN

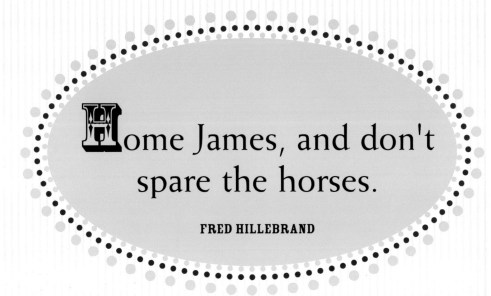

Home James, and don't spare the horses.

FRED HILLEBRAND

Thanks for the ride, Prince Charming.

CARRIE BRADSHAW, *SEX AND THE CITY*

Hey baby, you've got something on your butt: my eyes.

ANONYMOUS

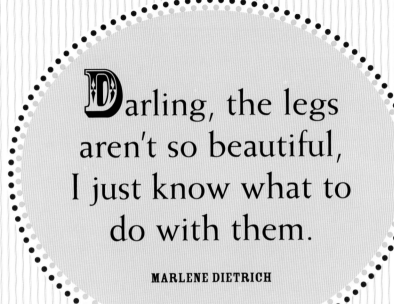

Darling, the legs aren't so beautiful, I just know what to do with them.

MARLENE DIETRICH

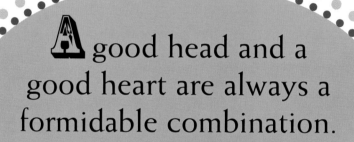

A good head and a good heart are always a formidable combination.

NELSON MANDELA

If a man does not make new acquaintances as he advances through life, he will soon find himself alone. A man should keep his friendships in constant repair.

SAMUEL JOHNSON

A birdie with a
yellow bill
Hopped upon the
window sill
Cocked his shining
eye and said:
'Ain't you shamed,
you sleepy head?'

R.L. STEVENSON

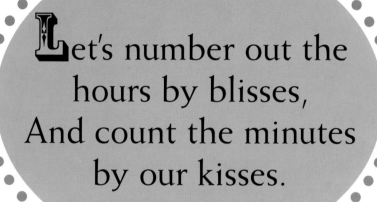

Let's number out the hours by blisses,
And count the minutes by our kisses.

JASPER MAYNE

It was a quiet way—
He asked if I was his—
I made no answer of
the Tongue
But answer of
the Eyes—

EMILY DICKINSON

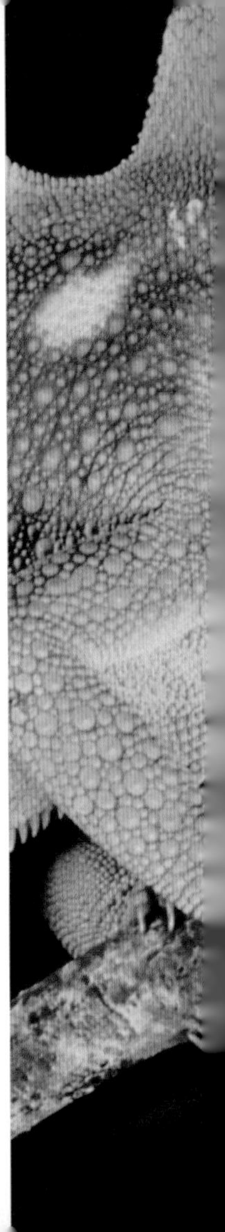

Often the difference between a successful marriage and a mediocre one consists of leaving about three or four things a day unsaid.

HARLAN MILLER

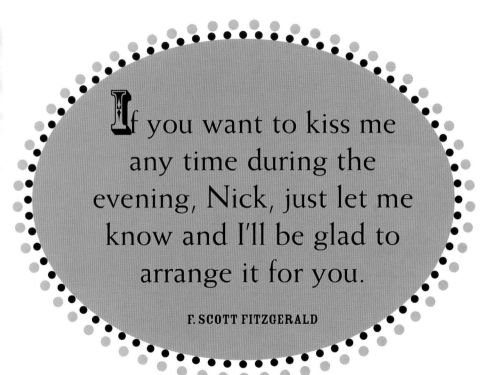

If you want to kiss me any time during the evening, Nick, just let me know and I'll be glad to arrange it for you.

F. SCOTT FITZGERALD

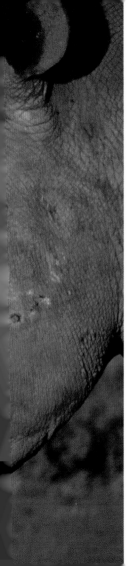

Every relationship I've been in, I've overwhelmed the girl. They just can't handle all the love.

JUSTIN TIMBERLAKE

> We're eyeball to eyeball, and I think the other fellow just blinked.
>
> **DEAN RUSK**

If I could rearrange the alphabet, I'd put U and I together.

ANONYMOUS

Ah, how good it feels! The hand of an old friend.

HENRY WADSWORTH LONGFELLOW

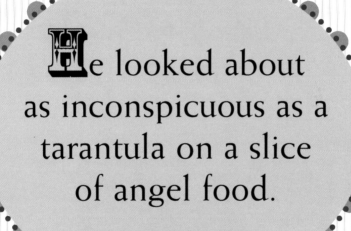

He looked about as inconspicuous as a tarantula on a slice of angel food.

RAYMOND CHANDLER

Don't get up. And please stop acting as if I were the Queen Mother!

BETTE DAVIS

I never found the companion that was so companionable as solitude.

HENRY DAVID THOREAU

> To catch a husband is an art; to hold him is a job.
>
> **SIMONE DE BEAUVOIR**

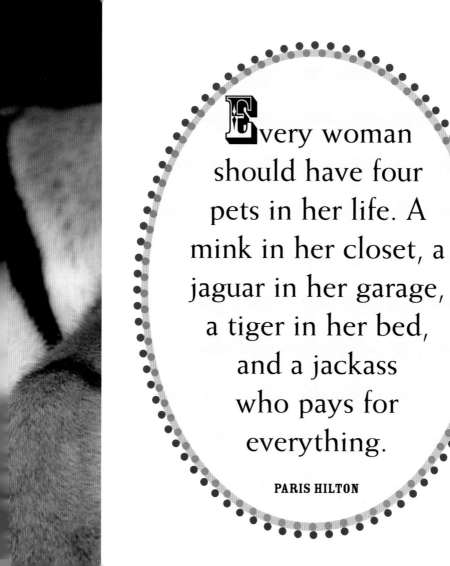

Every woman should have four pets in her life. A mink in her closet, a jaguar in her garage, a tiger in her bed, and a jackass who pays for everything.

PARIS HILTON

Don't talk to me about rules, dear. Wherever I stay I make the goddam rules.

MARIA CALLAS

He ain't heavy,
he's my brother.

BOBBY RUSSELL

Life is like a box of crayons. Most people are the 8-color boxes, but what you're really looking for are the 64-color boxes with sharpeners on the back.

JOHN MAYER

What does it matter that we take different roads so long as we reach the same goal?

GANDHI

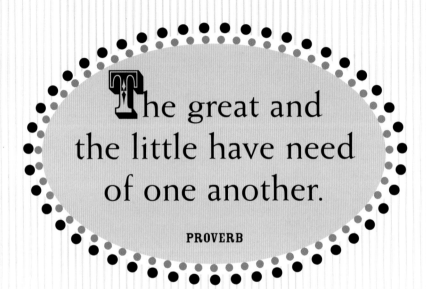

The great and the little have need of one another.

PROVERB

When buffeted and
beaten by life's storms,
When by the bitter
cares of life oppressed,
I want no surer haven
than your arms,
I want no sweeter
heaven than
your breast.

JAMES WELDON JOHNSON

This is a boy, sir. Not a girl. If you're baffled by the difference it might be as well to approach both with caution.

JOE ORTON

I do not know how to kiss, or I would kiss you. Where do the noses go?

INGRID BERGMAN

I was in a bar and I said to a friend, "You know, we've become those 40-year-old guys we used to look at and say, 'Isn't it sad?'"

GEORGE CLOONEY

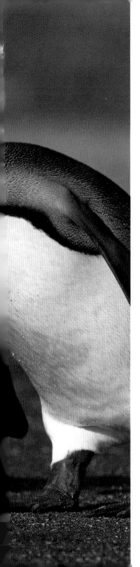

There is nothing more galling to angry people than the coolness of those on whom they wish to vent their spleen.

ALEXANDRE DUMAS

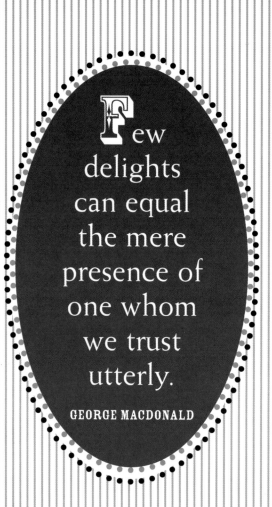

Few delights can equal the mere presence of one whom we trust utterly.

GEORGE MACDONALD

Do you believe in love at first sight, or should I walk by again?

ANONYMOUS

The two women exchanged the sort of glance women use when no knife is handy.

ELLERY QUEEN

Picture Credits

p.7 and p.95 © Tim Davis/CORBIS; p.8 © Renee Lynn/CORBIS; p.11 and p. 84 © Paul Kaye; Cordaiy Photo Library Ltd./CORBIS; p.12 © Lorne Resnick/Age Fotostock; p.15 © Julie Habel/CORBIS; p.16 and p. 31 Theo Allofs/CORBIS; p.19 Motor-Presse-Syndicat/Super Stock; p.20 © Gallo Images/CORBIS; p.23 ©DiMaggio/Kalish/CORBIS; p.24 and p. 44 © Rutz /Mauritius; p.27, p. 79 and p. 91 © Hulton-Deutsch Collection/CORBIS; p.28 Haruyoshi Yamaguchi/CORBIS; p.32, p.39, p.43, p.50, p.60 and p.71 © Bettmann/CORBIS; p.35 Chris Jones/CORBIS; p.36 and p.76 © Fritz Poelking/Age Fotostock; p.40 © Peter Greste /Reuters/CORBIS; p.42 © Paul A. Souders/CORBIS; p.47 and p.72 David Aubrey/CORBIS; p.48 © The Scotsman/CORBIS SYGMA; p.52 © Martin Harvey/CORBIS; p.55 © Morton Beebe/CORBIS; p.56 © Ilya Naymushin/Reuters/CORBIS; p.59 © Jonathan Blair/CORBIS; p.63 © D. Robert & Lorrie Franz/CORBIS; p.64 © Julie Habel/CORBIS; p.67 © Kit Kittle/CORBIS; p.68 © Sukree Sukplang/Reuters/CORBIS; p.75 © Cynthia Diane Pringle/CORBIS; p.80 and p.92 © Reuters/CORBIS; p.83 © DK Limited/CORBIS; p.87 © Mary Ann McDonald/CORBIS;

Text Credits

p.10 extract from *Is Marriage Lawful?* by A. P. Herbert (Methuen, 1935); p.13 dialogue from *Absolutely Fabulous* (BBC; screenplay Jennifer Saunders); p.51 extract from *The Great Gatsby* by F. Scott Fitzgerald (OUP, 1998); p. 61 extract from *Farewell My Lovely* by Raymond Chandler (Penguin, 1988); p.73 lyric from *He Ain't Heavy, He's My Brother* by Bobby Russell © EMI Records; p.82 extract from *What The Butler Saw* by Joe Orton (Methuen, 1969); p.85 dialogue from *For Whom The Bell Tolls* (Paramount Pictures/Emka Ltd; screenplay Dudley Nicholls);